Copyright ©2020 I RICHARDS

All rights reserved. No part of this publication may be reproduced, distributed, or transmitted in any form or by any means, including photocopying, recording, or other electronic or mechanical methods, without the prior written permission of the publisher, except in the case of brief quotations embodied in critical reviews and certain other noncommercial uses.

Table of Contents

Copyright ©2020 DR. JOHN RICHARDS 1

All rights reserved. No part of this publication may be reproduced, distributed, or transmitted in any form or by any means, including photocopying, recording, or other electronic or mechanical methods, without the prior written permission of the publisher, except in the case of brief quotations embodied in critical reviews and certain other noncommercial uses........... 1

INTRODUCTION 4

    Historical Note And Terminology 5

    Causes Of Essential Tremor 6

    Who Gets Essential Tremor 7

    Symptoms Of Essential Tremor 7

    Can Essential Tremor Increase The Risk For Other Illnesses ... 8

    How Essential Tremor Are Diagnosed 8

    How Is Essential Tremor Treated... 9

    Can Essential Tremor Be Prevented 10

    Can Essential Tremor Be Cured 10

    Pathogenesis Of Essential Tremor... 10

    Related Disorders 12

    Proven Ways To Treat Essential Tremor 14

    Herbal Remedies For Essential Tremors................................. 17

    Essential Oils For Essential Tremor... 19

    CBD For Essential Tremor 21

Essential Tremor And Natural Remedies 22
Conventional Treatment Of Essential Tremor 23
CBD Dosage And Use 25
Mediterranean Diet And Essential Tremor 26
Conclusion 28

INTRODUCTION

Essential tremor is the most common movement disorder in adults. Many people with essential tremor have not been diagnosed by a physician even though many report functional disability. This article discusses the clinical presentation, differential diagnosis, pathophysiology, and treatment of essential tremor. Publications pertaining to the diagnosis, genetics, and treatment of essential tremor are reviewed. There is an ongoing effort to update the classification of tremor, including a new tremor classification system, which may impact essential tremor clinical diagnosis. Unfortunately, available medications are frequently inadequate, and work on new therapeutics is limited. Thalamic deep brain stimulation is effective in most patients, and new stereotactic surgical sites are being explored. New surgical techniques are now available. There is considerable evidence that essential tremor is a clinically defined tremor syndrome with multiple etiologies; this probably explains much of the difficulty in identifying underlying causes and developing new treatments.

Essential tremor is a primary tremor syndrome

- The extent of nontremor symptoms possible within essential tremor plus is disputed but at least includes mild changes in balance, and in some patients, mood and cognitive changes.

• Parkinson disease, dystonia and other tremorogenic conditions are frequently misdiagnosed as essential tremor.

• Abnormal oscillation in the cerebellothalamocortical pathway likely produces tremor, but the structural, biochemical, and genetic mechanisms of essential tremor are uncertain.

• There is slow growth in tremor treatment options. Pharmacotherapeutic options (eg, primidone and propranolol) are limited and vary in efficacy. Occupational therapy and mechanical nonmedication interventions, such as specialized utensils, are increasing in variety and utility. Deep brain stimulation and unilateral standard and focused ultrasound thalamotomy surgeries are effective therapeutic options for patients with disabling tremor that responds inadequately to other interventions.

• Treatments focusing on nonmotor areas such as psychosocial coping, anxiety, and depression can improve overall quality of life as well as tremor treatment.

Historical Note And Terminology

Tremor is an involuntary rhythmic oscillatory movement of any body area. Some level of involuntary oscillation, called physiologic tremor, is normal. At the other end of the spectrum, pathologic states produce involuntary oscillations ranging from mild to disabling. Tremor is the most common neurologic movement disorder. Tremor is categorized as rest, postural, or kinetic tremor according to whether the tremor occurs in repose, steady posture, or movement. Descriptions of tremor date back centuries, to Ecclesiastes XII: and ancient India and Egypt . The specific distinction of kinetic tremor can be found in Galen's writings and many accounts since. Historically, the distinction between action tremor (postural tremor or kinetic tremor) and rest tremor was a seminal event in distinguishing essential tremor from Parkinson disease. The term "essential tremor" was developed in the 1800s to describe an isolated action, often familial, tremor.

Causes Of Essential Tremor

The true cause of Essential Tremor is still not understood, but it is thought that the abnormal electrical brain activity that causes tremor is processed through the thalamus. The thalamus is a structure deep in the brain that coordinates and controls muscle activity.

Genetics is responsible for causing ET in half of all people with the condition. A child born to a parent with ET will have up to a 50% chance of inheriting the responsible gene, but may never actually experience symptoms. Although ET is more common in the elderly and symptoms become more pronounced with age it is not a part of the natural aging process.

Who Gets Essential Tremor

Essential Tremor is the most common movement disorder, affecting up to 10 million people in the U.S.

While ET can occur at any age, it most often strikes for the first time during adolescence or in middle age (between ages 40 and 50).

Symptoms Of Essential Tremor

The primary symptoms associated with essential tremor include:

- Uncontrollable shaking that occurs for brief periods of time
- Shaking voice
- Nodding head
- Tremors that worsen during periods of emotional stress
- Tremors that get worse with purposeful movement
- Tremors that lessen with rest
- Balance problems (in rare cases)

The uncontrollable shaking associated with ET is not unique to this condition. Many different factors or diseases can also cause tremors, including Parkinson's disease, multiple sclerosis, fatigue after exercise, extreme emotional distress, brain tumors, some prescription drugs, metabolic abnormalities, and alcohol or drug withdrawal.

Can Essential Tremor Increase The Risk For Other Illnesses

Essential Tremor is linked to other illnesses. Other movement disorders, such as Parkinson's disease, have been associated with ET. Some reports have also linked ET with migraine headaches. People with ET may also be at a high risks of developing dementia (particularly Alzheimer's disease).

Drugs used to treat Essential Tremor may also increase a person's risk of becoming depressed.

Some experts also feel that there is no increased risk for Parkinson's disease for people with ET. Instead, some people diagnosed as having ET are initially incorrectly diagnosed and subsequently turn out to have Parkinson's disease.

How Essential Tremor Are Diagnosed

A neurologist or movement disorder specialist can usually diagnose Essential Tremor based on reported symptoms and a complete neurological exam. There is no specific blood, urine, or other test used to diagnose ET.

As part of the exam, your health care provider may consider other causes of tremor, such as thyroid disease, excessive caffeine intake, or medication side effects.

How Is Essential Tremor Treated

Mild essential tremor may not require treatment. However, if ET interferes with your ability to function or if you find it socially unacceptable, there are treatments that may improve symptoms. Treatments may include medications or surgery.

Medications: Oral drugs can significantly reduce the severity of essential tremor. Medications include propranolol (Inderal), primidone (Mysoline), gabapentin (Neurontin), and topiramate (Topamax). Other drug options include the benzodiazepines lorazepam (Ativan), clonazepam (Klonopin), diazepam (Valium), and alprazolam (Xanax). Botox injections may also be a treatment option. This treatment has been effective for vocal and head tremors.

Surgery: Deep brain stimulation (DBS) is a surgical treatment option for people with severe tremor despite medical therapy. DBS involves surgical implantation of electrical leads into the thalamus. This is an area deep within the brain that coordinates muscle control that is thought to be affected in ET.

MRI-guided Focused High Intensity Ultrasound: Neuravive uses magnetic resonance images (MRI) to focus ultrasound to destroy tissue in the thalamus. Patients are awake and responsive during the entire treatment.

Can Essential Tremor Be Prevented

Because we do not know the exact cause of Essential Tremor, there is currently no way to prevent it. However, knowing that ET has a genetic link brings us further in the search for effective treatments and, ultimately, ways to prevent it.

Can Essential Tremor Be Cured

There is no cure for essential tremor, but treatments that provide relief from its symptoms may be helpful in improving quality of life. These include drugs and surgery that ease tremor. But not every treatment or procedure is effective for every person with ET. Your doctor will recommend an individualized treatment plan, including certain lifestyle changes that may help to reduce your tremors.

Pathogenesis Of Essential Tremor

Essential tremor (ET) is the most common tremor disorder, twenty times more prevalent than Parkinson's disease. Up to 6% of the general population has ET. Uncontrollable trembling eventually forces 10 - 25% of patients to retire prematurely. There is no cure, and few medications lessen the tremor, although deep brain stimulation has provided promising results. Clinical evidence and neuro-imaging studies suggest that the cerebellum is centrally involved in ET, and evidence from clinical and animal studies suggests that there may be a disturbance in the gamma amino butyric acid (GABA) neurotransmitter system. While ET is clinically progressive, little is known about its underlying pathology. There have been few published postmortem examinations. The fundamental question in ET research is whether an underlying pathology can be identified in terms of morphological or morphometric changes of specific cell types in specific brain regions? Second, is there a neurotransmitter abnormality in ET, either resulting as a consequence of cell loss or in the absence of cell loss? The proposed study will be a collaborative effort involving four centers in the United States and Canada where archival postmortem tissue on 24 ET patients is available. In addition, with the help of the International Essential Tremor Foundation, we will establish at Columbia University a centralized repository for new prospectively-collected ET brains, collecting 36 additional ET brains during the five-year period. The 60 ET brains will be compared with 40 control brains. Primary Aim 1 is to study the pathology of ET to determine whether there are changes in specific brain regions. Using conventional morphological methods and quantitative morphometric assessments (stereology), tissue will be examined for changes, including cell loss, in the main region of interest (cerebellar hemispheres) and in secondary regions of interest (red nuclei, thalami, inferior olivary nuclei). We hypothesize that changes and cell loss in the cerebellum will be present to a greater extent in ET than in control brains. Primary Aim 2 is to study the GABA neurotransmitter system. We hypothesize that there will be differences in cerebellar GABA-ergic immuno-labeling in ET compared to control brains. Current therapies for ET have come to us by serendipity and are ineffective in up to 50% of patients. Knowledge of the pathological changes and neurochemical abnormality in ET is critical for the design of new therapies for ET.

Related Disorders

Symptoms of the following disorders can be similar to those of ET. Comparisons may be useful for a differential diagnosis.

Parkinson's disease (PD) is a slowly progressive neurologic movement disorder characterized by involuntary, resting tremor (trembling), muscular stiffness or lack of flexibility (rigidity), slowness of movement (bradykinesia) and difficulty controlling voluntary movements. Degenerative changes occur in areas deep within the brain (substantia nigra and other pigmented regions of the brain), resulting in decreasing levels of the neurotransmitter dopamine in the brain. Dopamine is a highly specialized brain chemical that sends a signal to other nerve cells, and participates in the regulation of body movements. Symptoms similar to those of PD (parkinsonian symptoms) may also develop secondary to hydrocephalus (a condition in which excessive cerebrospinal fluid accumulates the spaces in the brain [ventricles]. As a result, the fluid increases pressure in the brain, and the skull may become enlarged or bulge). Parkinsonian symptoms may also occur as a result of head trauma, inflammation of the brain (encephalitis), strokes (infarcts), or tumors deep within the cerebral hemispheres and base of the brain (i.e., basal ganglia), or exposure to certain drugs and toxins. PD usually begins in late adulthood. It is slowly progressive; however, it may not become incapacitating for many years.

Dystonia is a group of movement disorders that vary in their symptoms, causes, progression, and treatments. This group of neurological conditions is generally characterized by involuntary muscle contractions that force the body into abnormal, sometimes painful, movements and positions (postures). Dystonia may be focal (affecting an isolated body part), segmental (affecting adjacent body areas, or generalized (affecting many major muscle groups simultaneously). There are many different causes for dystonia. Genetic as well as non-genetic factors contribute to all forms of dystonia. The most characteristic finding associated with dystonia is twisting, repetitive movements that affect the neck, torso, limbs, eyes, face, vocal cords, and/or a combination of these muscle groups. Some individuals may have tremor as well. (For more information on this disorder, choose "dystonia" as your search term in the Rare Disease Database.)

There are many different potential causes of tremor. Such causes include the use of certain drugs including certain antidepressants; hyperthyroidism; and certain toxins including lead and mercury. Other forms of tremors exist as well, including psychogenic tremor, physiologic tremor, orthostatic tremor, and task-specific tremors such as primary writing tremor.

Proven Ways To Treat Essential Tremor

Are you among the 7 million Americans who have essential tremor? If your hand shakes involuntarily, or another part of your body trembles, you may have essential tremor.

Essential tremor (ET) can be confused with Parkinson's disease, but ET is different and much more common. A health care clinician can tell the difference between the two conditions.

Both women and men of any age can be affected by ET, but it's more common in middle age and later in life. The tremors can affect your arms, legs, trunk, head and even your voice.

We don't have a definitive cure for essential tremor. If the symptoms are mild, you may not even need a treatment. If the symptoms affect your daily activities, work or your quality of life, discuss your concerns with your health care clinician.

If the tremors are mild, some simple lifestyle changes may help:

1. Follow an appropriate sleep schedule. For some people, physical exhaustion can cause tremors.

2. Try relaxation techniques. This can work well for tremors brought on by stress or heightened emotions.

If the condition is more problematic, your clinician may do some tests to find the underlying cause. Depending on the cause, additional treatment options may be available:

3. Employ occupational therapy. An occupational therapist can help you adjust to living with the tremors. Some simple changes can make life easier. Use eating utensils with larger handles. Wearing wrist weights to stabilize the hand. Select clothes that are easy on and off — no buttons!

4. Avoid aggravating substances. Medications (like certain antidepressants, antiepileptics or inhalers) or foods (caffeine, energy drinks) can worsen tremor. Ask your doctor if any of your medications could be the source of your problems.

5. Take prescribed medications. Based on the underlying cause, we'll recommend a good option. A good result with medication would be reduction in tremor by about 50%.

Propranolol. This is a beta blocker. These meds are commonly used to treat high blood pressure. Don't use beta blockers if you have asthma or a heart problem. Side effects can include fatigue and lightheadedness.

Primidone. This anti-seizure drug is typically used to treat epileptic seizures. Side effects can include short-term drowsiness, concentration problems or nausea.

Botox. This injectable drug is an accepted treatment for conditions such as migraine, bladder dysfunction and excessive sweating. It can also be used to treat hand, head or voice tremors. When used for hand tremors, you may notice finger or wrist weakness. When used for voice tremors, Botox can cause a raspy voice or swallowing difficulties.

Various other medications can be tried including clonazepam, gabapentin, topiramate, zonisamide although these are generally less effective.

6. Utilize surgical treatments. These methods are used for bothersome or disabling tremor that is not adequately managed on medication. A good result with surgery would be elimination or near elimination of tremor.

Deep Brain Stimulation. This is the most effective and most proven tremor therapy and frequently can result in tremor freedom. A thin wire is surgically implanted into a deep region of the brain that is involved in generating tremor. Electrical stimulation delivered at the tip of the wire is adjusted by your doctor and powered by a battery pack placed in the chest. Deep Brain Stimulation offers the ability to treat tremor on both sides and can be adjusted over time. There are risks of bleeding, infection, speech or balance issues with the surgery.

Focused X-rays or Ultrasound. During stereotactic radiosurgery, the surgeon directs X-rays or Ultrasound at the specific part of the brain that's the source of the tremors. Special imaging technology helps direct the X-rays to the precise target. These techniques are limited to only treating one arm rather than both arms and cannot be adjusted after they are done.

Herbal Remedies For Essential Tremors

An essential tremor as a central nervous system disorder resulting in chronic shaking or rhythmic muscle spasms. These essential tremors may occur in nearly any portion of the body; however, the most common areas of shaking are isolated within the hands, head, arms or legs. The majority of essential tremors are caused by some form of genetic mutation, and scientists are still undecided as to what mutation causes this condition. Treatment plans for those with essential tremors may consist of synthetic prescription medications and possibly surgery; however, certain herbal remedies may be utilized upon the consent of your doctor to help reduce or relieve shaking.

- **Skull Cap**

Skullcap has been traditionally used as a mild relaxant, anti-anxiety herb as well as a natural remedy to help ease convulsions, such as those found with essential tremors. While current scientific research is inconclusive regarding its effectiveness for treating convulsions, due to the calming properties found within this herb it may be beneficial for those suffering with essential tumors. The American Botanical Council reports the King's American Dispensatory recommends this herb to help treat tremors and convulsions. Consuming 1 to 2 g of American skullcap per day or drinking two to three cups of freshly brewed skullcap tea per day is recommended.

- **Passionflower**

The use of passionflower for anxiety and seizures has been used in the Americas and throughout Europe, and this herb is commonly used for its relaxing qualities in modern day America. Scientists are unsure as to why this herb is beneficial for anxiety, but it is believed passionflower increases the levels of gamma-aminobutyric acid, or GABA, within the brain. By lowering brain GABA levels, brain cell activity is calmed, thus resulting in a relaxed state. It may be possible to reduce the severity of essential tremors by consuming this herb; however, the actual effectiveness of passionflower for tremors is unknown. Consuming this herb as a tea. To do so, pour 1 tsp. of dried passionflower into 8 oz. of boiling water and allow the tea to steep for 10 minutes. Strain and consume three to four servings per day is advisable.

- **Valerian**

Valerian has been used to treat nervous restlessness, anxiety and insomnia for thousands of years, and it continues to be among the most popular herbal remedies to promote relaxation. As with other calming herbs, scientists are unsure as to the actual reason why this herb works, but it is suggested it increase the GABA chemical in the brain, and the FDA recognizes valerian as generally safe for use. It is reported that valerian is commonly used to help treat mild tremors, thus it may be an effective natural remedy for tremors. Consume valerian tea by pouring 8 oz. of boiling water over 1 tsp. of dried valerian root and allow the tea to steep for 5 to 10 minutes. Consume up to three times per day.

- **Kava Kava**

Traditionally, kava kava was used as a ceremonial drink throughout the Pacific Islands; however, it is more commonly used as a relaxing agent. It is unclear as to why kava kava promotes a relaxed state; however, it is stated that this herb is capable of treating nervous disorders, anxiety and insomnia. While kava kava is possibly effective at treating essential tremors through its calming effects, users must be careful as this herb does carry a risk of liver damage. Only consume kava kava under the direct supervision of your healthcare provider.

Essential Oils For Essential Tremor

Essential Oils and Essential Tremor appear to have a positive link, and essential oils may help treat symptoms of Essential Tremor. Essential oils are a concentrated liquid containing aroma compounds from plants and other natural ingredients that have been found to help in certain symptoms of disease and illness. The compounds are made of small organic molecules that tend to change quickly from their liquid state, so it can be diffused and changed to move quickly in the air. Here have been over 3,000 varieties of these compounds found to date.

Such unique properties make essential oils ideal for aromatherapy. Aromatherapy has been proven to improve psychological and physical well being. It can be offered on it's own, or as a form of alternative medicine. Aromatherapy began back in 1937 for a range of ailments. Essential oils can be rubbed on the skin using a coconut or transfer oil, or some can be consumed in tea or water. Some essential oils are not consumption grade, so it is important to do your research before purchasing and using essential oils for consumption. Aromatherapy is not as intense, therefore may not have as many positive effects as consumption or rubbing the oil on the skin.

There have been certain types of oils that have been found to help manage symptoms of Essential Tremor. Some decide to use oils in conjunction with medication and other treatments, or use oils to completely manage their symptoms. The first oil that has been found to help with tremors is Frankincense Essential oil. This oil is generally used to relieve nervous energy and depressions. Once taken, this essential oil acts as a nervous system stimulant and can help to calm nervous tension.

The second most popular oil for treating Essential tremor is Vetiver Oil. Like Frankincense oil, it acts as a stimulant which helps calm the nervous symptoms, in which the tremors usually start from. The third oil is Arnica oil. This oil, combined with marjoram and rosemary, helps to reduce inflammation and muscle spasms. Lemon and orange essential oils both work with the body to help boost mobility. When using Arnica oil, it is usually used by mixing it with water and sipping. Doing this three times daily can help quake relief.

Helichrysum oil is another option for those looking for tremor relief. The properties in Helichrysum lower the level of brain inflammation, and because of Essential Tremors link to the brain, this may help to relieve some of the symptoms. Cinnamon Oil is generally used for treating diabetes and helping those with diabetic symptoms. These properties also make it a great product for Essential Tremor. Oils such as Cedarwood and Sandalwood are high in Sesquiterpene, which cross the brain blood barrier. They carry more oxygen throughout the body and help fire neurons and relieve neurological symptoms.

Overall, along with other treatments, Essential oils can greatly benefit those with Essential Tremors and help symptoms be more manageable in a natural and holistic way. Along with our product Tremor Miracle, Essential oils can be a great way to add a bit more comfort and relief to tremors and shaky muscles.

CBD For Essential Tremor

We have already talked about the healing properties of CBD when treating Multiple Sclerosis, Parkinson's Disease and substance withdrawal. But how are these conditions related to essential tremor and where does CBD come in?

It turns out that essential tremor could either be passed genetically from afflicted family members or could be a side-effect of these above-mentioned conditions.

As there has been medical proof that CBD can effectively treat Multiple Sclerosis, Parkinson's Disease and substance withdrawal, one can only hope that this will also be the case when it comes to essential tremor. In this article, we will look at the relationship between CBD and essential tremor.

CBD is a cannabinoid derived from the cannabis plant that has no psychotropic effect, also known as hemp. It does not cause the symptoms of dizziness or euphoria otherways associated with cannabis use. It has many therapeutic properties, low toxicity and its minimum, or in some cases none, side effects.

Essential Tremor And Natural Remedies

Essential tremors usually occur in the hands, presenting in up and down movements. However, they can also affect the arms, head, face, and larynx (voice box), which results in a shaky voice. People with essential tremors may remain undiagnosed for a longer time, as at first, the symptoms are rather discreet and can be easily mistaken with ordinary tremor. Ordinary tremor affects most people and is considered normal, especially after the use of caffeine and cigarettes.

Tremors can range from minor to severe. Minor tremors may not affect your everyday life, but some severe cases can seriously interfere with a person's daily activities.

As mentioned before, tremors can occur in different body parts, therefore the symptoms can vary. The most common symptoms of essential tremor go as follows:

Noticeable shakiness in the hands or arms when writing, holding thing or using your hands in general;

Involuntary movements of the head and neck such as shaking in an up-and-down or side-to-side motion;

Twitchy eyelids;

Tremors in the tongue or voice box make a person's voice sound shaky when speaking;

Abnormal walking or loss of balance are common for tremors in a person's core, legs, and feet;

Essential tremor usually affects people over the age of 40. Nonetheless, genetics also play a crucial role when it comes to this condition, as essential tremor can also be inherited. If essential tremor runs in the family, it is referred to as " familial tremor". Parents who have developed this condition carry a 50% chance of passing it to their children.

Conventional Treatment Of Essential Tremor

There is no known cure for essential tremor, however, doctors offer several treatments to keep the condition at bay. These include anti-seizure drugs, brain surgery, also known as deep brain stimulation (DBS) and ultrasound MRI treatment, aiming to destroy tissue in the thalamus.

All these treatments have proven some success, however, they are quite invasive and expensive.

CBD and essential tremor – an alternative treatment CBD oil for essential tremor can provide an alternative and less invasive treatment of this condition. Medical research has already extensively looked into cannabinoids and their ability to relax the muscles and calm seizure.

Perhaps the most powerful property of CBD is its direct connection to the endocannabinoid system (ECS). ECS and its endocannabinoid receptors are the main regulators of our bodily functions such as pain, hunger, emotions etc., which also means that when there is an internal issue, it is up to that system to initiate the healing process. Consuming CBD increases the production of endocannabinoids in the body, which enhances the healing process.

When using CBD oil for essential tremor the cannabinoids in the oil can target and regulate the ESC receptors responsible for the muscle movement. These are located in both the brain and throughout the body. By regulating these receptors CBD oil enables relaxation, reduce inflammation, and ease symptoms of tremors. There have been studies that confirm these interactions and confirm that CBD can effectively act as a neuroprotectant.

The already existing research has sparked a lot of interest recently and CBD may be on its way of becoming an official treatment for essential tremor.

This is because research has officially started a trial on the effects of CBD and THC on essential tremor. The trial has been FDA approved and is expected to release the result as the end of 2019. The lead investigator of this clinical trial, scientists are very passionate about finding an adequate treatment for essential tremor in the face of CBD. Essential tremor is 10 times more common than Parkinson's and yet nobody really knows about essential tremor. That we're finally getting to a potential therapeutic option in an area that is untapped is a big deal.

CBD Dosage And Use

There are two ways to take CBD for essential tremor. As everyone's reaction to CBD is different, we encourage people to experiment unit they find the one that gives them the best essential tremor relief.

- **Sublingual**

CBD oil comes in small bottles with a dropper. A couple of drops are put under the tongue and then the oil is absorbed into the body.

- **Vaping CBD E-Liquids**

A faster way to feel the relief and perhaps one of the most popular natural remedies for essential tremor is to vape CBD E-liquids. The vapor enters the lungs and speedily reaches the bloodstream, providing a calming (not psychoactive) effect

When it comes to dosage, there is no one-size-fits-all method. As mentioned before, everyone's reaction to CBD is different, so you have to listen to your body and take notice of improvement.

The approach suggest to start low, such as 1 drop, 3 times a day and gradually increase the dosage every 3 days, until you reach 3 drops, 3 times a day. On average, 25mg a pay is a sufficient amount for someone with non-severe symptoms.

There are many testimonials online regarding the effectiveness of CBD over essential tremor. A CBD shop recently published the story of a customer, expressing their gratitude and revealing the positive effect CBD products have had on her conditoin. The woman claims her struggles with essential tremor and porphyria have been eliminated thanks to CBD products. CBD for essential tremor can be effective. The "twitching" motions can be greatly reduced, resulting in normal daily activities. Whether the oil is taken orally, or an e-liquid is vaped, there is evidence that CBD works well for a large number of people. As a much safer and inexpensive (in comparison to other drug treatment and surgery), it is an effective way to improve day-to-day life. Essential tremor does not always need treatment. If the condition isn't seriously impacting your life regularly, then alternative action might be all that is required to manage the symptoms.

Mediterranean Diet And Essential Tremor

"Eat an apple on going to bed, and you'll keep the doctor from earning his bread." The first known use of this adage occurred in Wales in the 1860s, but apples have a respectable history of being recognized as a support for healthful living. While some foods can indeed influence how healthy a person is, healthy nutrition is more complex than any single food item.

In fact, the whole contribution of food to health may be greater than the sum of the individual ingredients. It's probable that a synergistic effect occurs with many culturally-based diets. For example, the Mediterranean Diet is made up of individual foods, each of which is generally believed to be healthy:

- Vegetable, legumes, fruits
- Cereals (preferably whole grain)
- Fish
- Monounsaturated fatty acids
- Low levels of dairy, meat and poultry
- Low to moderate alcohol consumption

By some mechanism as yet not fully understood, when taken together consistently over time, it appears that the Mediterranean diet (MeDi) enhances cardiovascular health, reduces inflammation, and protects against degeneration related to oxidation. In particular, consuming less inflammatory foods (many dairy products contribute to an inflammatory response in the body) and keeping tissues in more youthful tone with protective antioxidants are recognized ways to minimize risk of cancer, diabetes and heart disease. Even the probability of developing Alzheimer's has been lowered among adults who live by the MeDi.

This raises the possibility that the MeDi helps cushion the body against neurodegeneration, a definite factor in Alzheimer's, and considered by some to be involved in Essential Tremor (ET). In a 2007 case study, it is hypothesized that greater adherence to the MeDi would be linked with lower odds of developing ETDi

The research involved 398 individuals (148 ET patients and 250 matched controls). The authors gathered data on the nutrition/diet of each participant. They found that those who more strictly followed the MeDi had "significantly lower odds of ET." Compared with those who had the least MeDi consumption, the intermediate group had 59% less risk, and those in the strictest group had 71% reduction.

Aside from the fact that there is a genetic component (ET tends to run in families), there also appears to be nongenetic influences (diet and toxins) that may have a role in developing ET. The authors wrote, "Consumption of many nutritional antioxidants did not differ between ET patients and controls according to a previous study, but the MeDi, being a composite dietary pattern, may be better at capturing the overall antioxidant effect of the diet." This speaks to the idea that a total and/or synergistic effect of the MeDi is perhaps more powerful than taking a handful of antioxidant supplements every day.

In other words, the ancient wisdom of Hippocrates may well apply to preventing ET: "Let food be thy medicine, and medicine be they food."

Conclusion

Although sometimes daunting, the patient with tremor offers the clinician the opportunity to demonstrate clinical expertise. Focusing on basic phenomenology, most patients can be diagnosed without ancillary testing, and treatment can be initiated quickly. Natural remedies could help relieve stress and pains that accompany surgical treatment of essential tremors. Less expensive and easily accessible to patients.

Printed in Great Britain
by Amazon